THE BIRDS OF KAUA'I

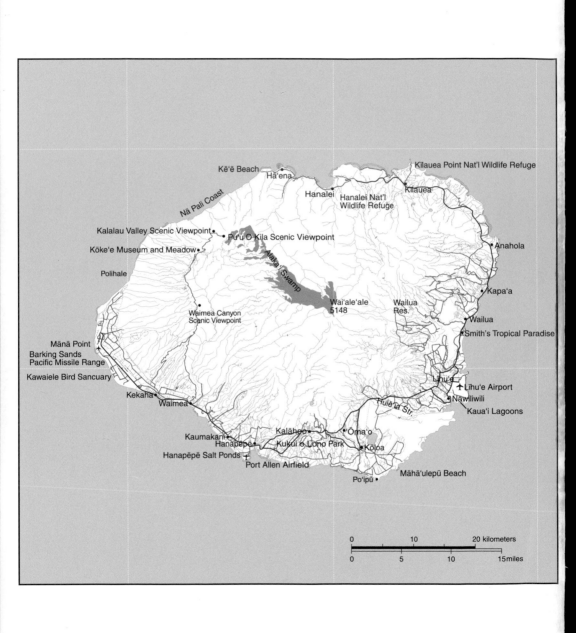

The Birds
of Kaua'i

JIM DENNY PHOTOGRAPHS BY THE AUTHOR

 A LATITUDE 20 BOOK University of Hawai'i Press, Honolulu

04 03 02 01 00 99   5 4 3 2 1

Library of Congress Cataloging-in-Publication Data

Denny, Jim, 1946–
  The birds of Kaua'i / Jim Denny ; photographs by the author.
    p.   cm.
  "A Latitude 20 book."
  Includes bibliographical references (p.   ) and index.
  ISBN 0–8248–2097–5 (alk. paper)
  1. Birds—Hawaii—Kauai.   I. Title.
QL684.H3D45   1999
598'.09969'41—dc21   98–39232
                                                          CIP

Book design by Kenneth Miyamoto
Cover design by Santos Barbasa

Printed by Friesens Corporation

To my father

At his side, I learned to love the natural world

Contents

Preface

IN THE LAST few years, several knowledgeably written and beautifully illustrated field guides have been published about the birds of Hawai'i. However, not every bird can be found on every island. It can be frustrating to have to page through all of the many species that exist throughout the Hawaiian Archipelago in an attempt to identify a bird seen on Kaua'i. The intent of this publication is to simplify your search. Every bird illustrated in this book has been seen in the backyards, parks, pastures, fields, taro patches, and canyons or along the seashores, roadsides, and forest trails of the Garden Island.

The species described in this guide are classified as endemic, indigenous, introduced, or migratory. Endemic species are those that evolved in the Islands and that reside naturally nowhere else. Indigenous species are those that arrived unaided by humans but are also known to reside in other locations. An indigenous species breeds here, but did not evolve here. The term native is often used to describe endemic as well as indigenous birds. Introduced species are those that arrived in the Islands aided by humans either accidentally or intentionally. The terms nonnative, alien, and exotic are often used to describe introduced species. Migratory species are those birds that breed elsewhere and visit Kaua'i only during certain seasons.

The scientific, Hawaiian, and English names used in this book are derived from the Checklist of the Birds of Hawaii—1997 by Robert L. Pyle, in 'Elepaio, the Journal of the Hawai'i Audubon Society (1997).

The Hawaiian names for the birds and plants as they appear here incorporate the pronunciation guides used in the *Hawaiian Dictionary* by Mary Kawena Pukui and Samuel H. Elbert (1986). Locations are spelled as described in *Place Names of Hawaii* by Pukui, Elbert, and Esther T. Mookini (1974). The *kahakō* or macron (¯) denotes a stressed vowel, and the *'okina* (') marks a glottal stop. English names for

The Alaka'i swamp is the most pristine native forest in Hawai'i. The isolated ridges and valleys of this preserve are the last remaining sanctuary for several species of critically endangered forest birds.

Hawaiian birds are listed where possible, and a scientific (Latin) name is given for every species.

The most treasured of the eighty-eight species listed in this guide are the native forest birds. These are discussed in their own section. Introduced species are also treated separately. All others are assembled into one of two groups—seabirds or wetland/shore birds. Those designated as endangered by the U.S. Fish and Wildlife Service are so noted.

Although the bulk of the information in the text is from my own observations, also included is a substantial amount of material collected by others. The serious observer or amateur naturalist who yearns for more information about the flora and fauna of Hawai'i can find a wealth of data among the references listed at the end of this book.

Finally a warning: Logic suggests that no one can get lost on an island, that if you were just to keep walking in a straight line, you would eventually meet the seashore. This logic does not apply to the ruggedness of Kaua'i's interior. The Alaka'i is so thick with vegetation and so deeply cut by the many valleys that drain it that it is impossible to walk in the same direction for any great distance. It is a dangerous place to go. It is tragic that people have perished in this Eden-like paradise. Stay on established trails and dress to stay dry. Ethnographers interpret the Hawaiian phrase *aloha 'āina* as a very old concept meaning a deep love for the land. As you seek to find the birds depicted in this book, respect this land.

Acknowledgments

MANY PEOPLE have helped make this book possible. I am indebted to field researcher Tom Snetsinger for his help with early versions of the manuscript and for sharing with me his vast knowledge of birds. I am grateful to noted wildlife photographer Jack Jeffrey for teaching me much about the craft of capturing a bird on film. When I am old and feeble, I will look back with fondness on the many hours spent searching for birds with Tom or photographing with Jack in the remote valleys of the Alaka'i. For his many inspiring articles about Kaua'i's native flora and fauna, I thank educator and naturalist David Boynton. His work continues to impress upon the public the uniqueness of the Hawaiian ecosystem. I thank Tom Telfer at the Kaua'i Division of Forestry and Wildlife for sharing with me his encounters with the nearly extinct 'Ō'ū and *Nukupu'u*, for his guidance in locating some of the more illusive lowland game birds, and for his excellent suggestions for improving the manuscript. I am indebted to Dr. Carter Atkinson and his assistant Julie Lease of the U.S. Geological Survey, Biological Resources Division, for allowing me to accompany them on their mist-netting trips to study the spread of avian malaria in our endemic forest birds. I am grateful to Capt. Tom Daniels, commander of the Navy's Pacific Missile Range at Barking Sands, for granting me access to the base to photograph the Laysan Albatross. My thanks to the dedicated participants of the Kaua'i forest-bird survey team (Thomas Kaiakapu, Paul Conry, Thane Pratt, and Tonnie Casey) for allowing a newcomer into their midst. I gained from them much more than I contributed. My gratitude to Dr. Bob Pyle of the Bishop Museum for permitting me to view the museum's collection of preserved skins of Hawai'i's extirpated birds and for guiding me through some of the finer points of taxonomy. My appreciation to Dr. Storrs Olson and Dr. David Burney of the Smithsonian Institution for allowing me onto their excavation site at Māhā'ulepū to photograph the fossil bones of Kaua'i's long-extinct birds. For sharing with me his exten-

Kaua'i *'Elepaio* on *'ōhi'a lehua.*

sive knowledge of game-bird distribution on Kaua'i, I am indebted to game-bird breeder Dean Nadatani. My thanks also to Bernie Agor Jr. for the gracious loan of his beautifully stuffed pheasant, which appears regally posed in the grass in the photo in the section on introduced birds. I owe the same gratitude to George Coates for the handsome Black Francolin that appears in the photo of that species. Thanks to Dekalb Hanapēpē Corn Research for allowing me to photograph the Rose-ringed Parakeet in their fields. I also appreciate the assistance of Laura Ishii of the State Department of Agriculture in helping me understand the regulations regarding the importation of alien species to Kaua'i. My thanks to *Puaiohi* researcher Christina Herrmann for her valuable advice as to how to improve this text. I am also grateful to Dr. Sheila Conant, of the University of Hawai'i, Department of Zoology. Her review of the manuscript led to several excellent suggestions that have vastly improved the scholarship of this work. I would like to express my appreciation to Hawaiian language instructor Paul Williams for his help in translating some of the Hawaiian names and phrases used in this text, and to Patricia Crosby, Keith Leber, and Masako Ikeda at University of Hawai'i Press and Eileen D'Araujo, for their patience and valuable assistance in getting through the publishing process. Finally, I wish to thank my family, who endured my obsession to finish this project. *Mahalo.*

Introduction

'O ka manu 'Ō'ō i mālama,
A he nani kou hulu ke lei 'ia.

O precious *'Ō'ō* bird,
Your feathers are so beautiful
   woven into a *lei.*

Mūkīkī ana 'oe i ka pua lehua
Kāhea ana 'oe i ka nui manu.

You sip *lehua* flowers
And call other birds.
   (Emerson 1909)

WE CAN ONLY imagine what birds the sea-weary Polynesians encountered when they first stepped ashore on Kaua'i. The generation of individuals that occupied that initial voyaging canoe has long since vanished, leaving no written record of the event. However, from the few chants that have survived, we know that the early inhabitants were absolutely filled with awe at the majestic beauty of this land and at the diversity of life upon these shores.

The ability of Kaua'i's native birds to inspire wonder has not yet vanished. In the forested uplands, some of the species seen by those first humans continue to thrive. From high in the dark green canopy, where the rain clings to the scarlet blossoms of the *'ōhi'a lehua,* to the forest floor where the fragrance of *mokihana* punctuates the crisp, cold air, the Kaua'i of old remains relatively untouched. *Hāpu'u* ferns grow as big as trees; *maile lau li'i,* the small-leaved vine cherished by *lei* makers throughout Hawai'i, tangles its way within the understory; and towering lobelias catch your attention as they reach up through the lushness to claim their place in the light. Each morning in this ancient forest a spectacle of sound and color that has welcomed the Alaka'i dawn for thousands of years begins anew. The earth comes alive with thousands of multihued birds moving among the branches of native plants in search of nourishment.

Among this colorful assembly is a distinct subfamily of birds known as Hawaiian honeycreepers. The adaptations of these birds are without equal in the animal kingdom. What so intrigues those who study them is the conjecture that all the members of this subfamily (more than fifty distinct species and subspecies have been identified throughout Hawai'i to date) evolved from a single species of finchlike ancestor—possibly a

The red blossoms of the *'ōhi'a lehua* are a major source of nectar for many native birds. Curved-billed honeycreepers also feed on the tubular blossoms of lobelias like this towering *hāhā lua* reaching up for the light in the Kaua'i rain forest.

small flock or even one lone pair of birds. It is not known how the first individuals arrived. Hawai'i is but a speck in the middle of a vast ocean, a huge distance for small songbirds to cross. Perhaps they were carried here by a mighty hurricane. It is interesting to reflect how a wind like Hurricane 'Iniki, which devastated the island of Kaua'i in 1992, might at a time in the far distant past have delivered life.

However the founding species came, the subsequent population explosion eventually overwhelmed the food supply. Isolation, coupled with intense competition over a very long period slowly "selected" those accidentally different individuals that had an advantage—a slightly longer bill for probing deeper into leaf bases to feed on hidden insects or a slightly stronger mandible with which to open tougher seeds. By the time humans arrived, one finchlike species had evolved into many, each occupying separate niches within the ecosystem—a process known to biologists as "adaptive radiation." On Kaua'i, an island of only 553 square miles (1,433 km^2), the environment varies from the desertlike conditions of Barking Sands on the leeward side to the wettest place on earth at Wai'ale'ale. With so many different exploitable resources available to them, the Hawaiian honeycreepers "radiated" into an impressive array of colors, bill shapes, and feeding preferences. The red 'Apapane with its short bill feeds mainly in the tops of the trees taking nectar from the 'ōhi'a lehua, the green 'Akeke'e occasionally feeds on flowers but most often uses its twisted beak to force apart closely spaced leaves to remove the insects that lie within. The 'Akikiki, an uncommon, small, gray bird that is only found on Kaua'i, rarely takes nectar, opting to spend most of its time creeping along the branches and trunks of trees probing the moss for insects. In some, the adaptations are extreme. The 'I'iwi developed an incredibly long, curved bill and tubelike tongue to facilitate the harvesting of nectar from the many similarly shaped flowers in the forest.

As you seek out the birds in this book, your first objective should be to learn to distinguish native birds from introduced birds. When John Ledyard, an American traveling with Captain Cook, hiked into the mountain forests of Hawai'i, he remarked that the birds he saw there had "the liveliest and most variegated plumage than (sic) any I have ever met with" (Foster 1993). With more than 160 bird introductions to our state in the last 200 years, the plumage of Hawai'i's birds is even more varied than in 1778. Some of these introductions have been here for so long that we have come to think of them as native species. The Common Myna, plentiful along our roadways, predates Lili'uokalani, the last ruler of the Hawaiian Kingdom. Visitors from the North American continent can easily find the birds of their homeland. The Northern Cardinal, House Finch, and the Northern Mockingbird are all

Isolated in midocean, Hawaiʻi's plants and birds evolved a symbiotic relationship. Flowers like that of the nearly extinct *hau hele ʻula* provided nectar for honeycreepers. The birds played the role of bees and pollinated the flower.

here. Travelers from Asia can see the *Mejiro, Hwamei,* and the Spotted Dove. From South America comes the beautiful Red-crested Cardinal, and from Indonesia, the White-rumped Shama.

In the 1,500 years since that first landing, many changes have occurred. More than half of the astounding diversity of birds found in

Hawai'i by the voyaging Polynesians are now gone. It is sad that the majority of birds listed on the Federal Register of Endangered Species are Hawaiian. In the deep sediment of an ancient lake bed on Kaua'i's south shore, researchers have found the bones of an impressive number of long-extinct birds that once existed in our coastal areas. Some, like

Excavation of an ancient lake bed at Māhā'ulepū.

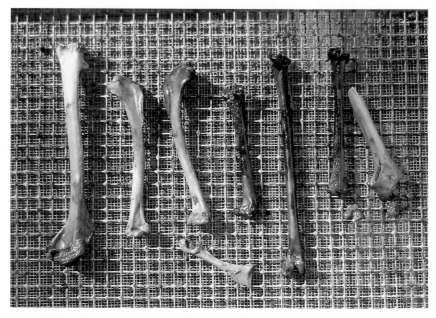

Fossil bones of long-extinct Kaua'i birds.

The Kōke'e State Forest, home to Kaua'i's native birds, flourishes right up to the rim of beautiful Kalalau Valley.

the *ʻIo (Buteo solitarius)* or Hawaiian Hawk, are known historically only from other islands. The sediment also contains the remains of many unknown species that were never recorded in the historical record on any island. These fossil bones give us a glimpse into the past at the birds those first voyagers may have seen. They include two rails, a long-legged owl, three geese (one much larger than the *Nēnē*), two Laysan-type finches, two Kona Grosbeak-type finches, at least one new species of a cone-billed finch, and three species of robust, flightless ducks known as *Moa nalo*. This latter group is an unusual new genus described as goose-size birds with massive hind legs, tiny wings, and bills like those of a tortoise.

Hawaiʻi's biota is not only famous for its wonderful adaptations but also for the numbers of its extinctions. The Kauaʻi *ʻŌʻō*, cherished by the Hawaiians for its beautiful feathers, probably no longer sips the *lehua* flowers in the forests of Kauaʻi, but some of the birds to which it called can still be seen. Perhaps it is time to take a closer look.

Native Forest Birds

A CENTURY AGO, the population of Kaua'i's native birds was so large that it appeared the island could not hold them all. Mrs. Francis Sinclair, a resident on Ni'ihau, an island 17 miles (27.3 km) off Kaua'i's leeward coast, remarked that after stormy weather, large numbers of Kaua'i's forest birds would be found washed up on the shores of that island. Fossil records indicate that they flourished from mountain to seashore. As the twentieth century drew near, however, a mysterious, widespread pandemic afflicted the once great Hawaiian flocks to the point that within a decade few birds could be seen in the lowlands. By the 1920s many of Hawai'i's birds had ceased to exist.

For reasons not clearly understood, the birds of Kaua'i escaped the full impact of this calamity. Remarkably, all fourteen species of endemic forest birds that were present at the time of Cook's arrival found a refuge in the remote valleys of the Alaka'i Swamp. In recent years, however, the Alaka'i has not proved to be remote enough. Several forest bird species that had sustainable populations twenty years ago have become extremely rare. No confirmed sightings of the *Kāma'o, Nukupu'u,* and *'Ō'ū* have been reported in the last decade. These species have declined to only a few individuals and are not expected to survive. The *'Akialoa* has not been seen since 1965 and is probably extinct. The *'Ō'ō,* if it still exists, has not been seen since the 1980s. Kaua'i now shares the grief of her sister islands and mourns for her native birds.

What has led to the loss of such a natural treasure? The causes are complex and not easily solved. Part of the problem lies with pigs, cats, and rats. Feral pigs are everywhere in the native forest. Conservationists have labeled them "Hawai'i's Forest Enemy #1."

When the pigs root up the forest floor, introduced weeds take hold, reducing native habitat. Their constant tilling increases erosion and creates breeding areas for mosquitoes. Why not just eradicate them? Brought to the Islands over a thousand years ago as a food source by

Brought by the colonizing Polynesians to sustain the lives of those who settled here, the feral pig is now "Hawai'i's Forest Enemy #1."

Hawai'i's pre-European discoverers, pig hunting is still an activity that is deeply rooted in local culture. Any proposal that would completely eliminate pigs from Hawai'i's native forest would be an emotionally charged issue certain to be unpopular among those who hunt them to feed their families.

Feral cats are a severe threat to native birds. Hurricanes 'Iwa in 1982 and 'Iniki in 1992 separated many cats from their homes, forcing them to fend for themselves. The progeny of these orphans have become a large, wild, breeding population of predators. These, in addition to unwanted cats intentionally released by their owners, roam the forest where they prey upon ground-nesting birds as well as climb trees to eat the fledglings of all species. It would be a difficult, expensive process to remove these wild creatures, which would be quickly replenished. Cats are not in short supply on Kaua'i.

The Polynesian Rat *(Rattus exulans)* is thought to have been introduced accidentally to Hawai'i as a stowaway aboard the same voyaging canoes that brought the Hawaiians. It and Europe's contribution, the Black Rat *(Rattus rattus),* are both agile climbers that can get to even high nests.

Rats are a major problem in the native forest. These introduced rodents climb trees during the night to eat the eggs of nesting birds.

A notable 1996 study by researchers Steve Fancy and Bethany Woodworth validated what many biologists had long suspected: the impact of these predatory mammals on forest bird populations is enormous. Their statistics revealed that the survival of 'Apapane chicks in a rat-free study area was more than twice that in an uncontrolled area. Survival of 'Elepaio chicks was almost four times greater! Why not just broadcast rat poison from a helicopter and kill them all? Pig hunters worry that such indiscriminate poisoning may also affect pigs and the dogs that hunt them.

Scientists suspect, however, that the biggest threat to our native forest birds is mosquito-borne disease. Until 1826, Hawai'i had no mosquitoes. What a paradise indeed! It is said that a whaling vessel in the process of refilling its freshwater casks dumped mosquito-infested water into a river on Maui. In the years that followed, the mosquito population exploded—spreading through all of Hawai'i.

The night-biting mosquito brought to Hawai'i by those sailors is *Culex quinquefasciatus,* a species that feeds on the blood of birds, biting them around the eyes, legs, and feet. The bite itself does little harm, but the

Avian disease researcher Dr. Carter Atkinson checks a mosquito trap in the Alakaʻi swamp.

mosquito can infect the bird with the parasite *Plasmodium relictum capristranoae*. The parasite invades the liver and spleen. It also penetrates the red blood cells, where it multiplies, ruptures the cell, and causes a debilitating anemic condition called avian malaria.

Although the *ʻApapane* demonstrates some resistance to this disease, others, like the *ʻIʻiwi*, have no immunity and are extremely susceptible. Recent studies by avian disease scientist Dr. Carter Atkinson have established that a single bite by one infected mosquito is nearly always fatal to this beautiful scarlet honeycreeper. Mosquitoes now reside in the upper reaches of the swamp where there were none before. It is hoped that a solution can be found before more of this island's native birds succumb to this virulent disease.

The Pihea boardwalk is the best place in Hawaiʻi to see forest birds. Eight of Kauaʻi's fourteen species of endemic forest birds as well as several introduced species can be found along this trail.

Not all of our native species are confined to the drawers of museum collections or to the yellowing pages of the reference library. Kaua'i still has much to offer. The Hawaiian word *pihea* means a loud din. Although it does not have the tumultuous uproar of birds it had centuries ago, the Pihea trail remains one of the best places in Hawai'i to see our endemic species. Hiking into the forest in search of native flora and fauna once meant standing ankle deep in mud in an often chilling rain. The chilling rain is still there, but the recent construction of a boardwalk by the State Department of Land and Natural Resources along the Pihea and the Alaka'i Swamp trails has made the search a much more pleasant experience.

'Akeke'e

Loxops caeruleirostris   4.5 inches (11 cm)

Referred to by some as the Kaua'i *'Ākepa* or *'Ō'ū holo wai*, this endemic species feeds high in the *'ōhi'a* canopy where it forages through the buds and terminal leaf clusters in search of insects. Its scissorlike bill is uniquely adapted to open the closely spaced leaves to achieve this purpose. Like the *'Akikiki*, the *'Akeke'e* travels in small family groups that call back and forth to each other as they feed. It is a busy bird, constantly moving in and out of the leaves. The Hawaiian word *'ākepa* literally means active, and this species certainly fits that description. Although the *'Akeke'e* is most common in the more remote Alaka'i, it is still present in the Kōke'e region and can be located without too much effort. It can frequently be seen near the intersection of the Pihea boardwalk and the Alaka'i Swamp trail.

'Akeke'e (Kaua'i *'Ākepa*), *Loxops caeruleirostris.*

'Akikiki

Oreomystis bairdi   5 inches (13 cm)

The Kaua'i Creeper gets its name from its habit of creeping along the trunks and limbs of trees, where it probes crevices in the bark for hidden insects. The *'Akikiki* is found only on Kaua'i and is formally being considered for listing as a threatened species by the U.S. Fish and Wildlife Service. Although this small, gray honeycreeper can occasionally be seen in the Kōke'e area, it is more often encountered in the Alaka'i Swamp. This pattern of retreat into the interior of the island concerns researchers, who consider such population trends a prelude to extinction. The *'Akikiki* travels in pairs or in small groups, the individuals of which call to each other as they feed. The call is a very short "sweet" and resembles closely the call of the *'Akeke'e*. Turn-of-the-century collectors found that the now critically endangered *Nukupu'u* sometimes kept company with the *'Akikiki*, so it is a good idea to search adjacent trees when the Kaua'i Creeper is sighted. The Mōhihi-Wai'alae ridge trail is a good place to see this species; however, if a four-wheel-drive vehicle and a full day are not available, the Alaka'i Swamp trail as it descends to Kawaikōī Stream has been a reliable stretch in recent years.

'Akikiki (Kaua'i Creeper), *Oreomystis bairdi.*

ʻAnianiau

Hemignathus parvus   4 inches (10 cm)

Found only on Kauaʻi, the *ʻAnianiau* is the smallest of all Hawaiian honeycreepers. In older literature it is sometimes referred to as the "lesser *ʻAmakihi*." Males are bright yellow and females somewhat duller in color. The *ʻAnianiau* is one of the most common native birds encountered along Kōkeʻe forest trails. Its short bill is adapted to feed on the flowers produced by such endemic plants as the *ʻōhelo, kanawao,* and *naupaka kuahiwi. ʻAnianiau* often feed in a circuit, so if one is patient enough it will return to the same plant and even the same flower. This small species can be readily identified by its call—a soft, slurred, two-note whistle rising in pitch. It is also helpful to remember that it is the only yellow-green honeycreeper without black on the face. Although common at this writing, the *ʻAnianiau,* like the rest of its honeycreeper cousins, is apparently susceptible to introduced diseases because it is rarely found below 3,000 ft (900 m), where mosquitoes are more common.

ʻAnianiau, Hemignathus parvus.

'Apapane

Himatione sanguinea sanguinea   5 inches (13 cm)

The *'Apapane* is Hawai'i's most common native forest bird. It is most often seen flying high above the Kōke'e forest canopy in search of the brilliant red *lehua* flowers of the *'ōhi'a*. This Hawaiian honeycreeper is also a gifted singer. Biologists who have ventured into the forest to study this beautiful endemic species are impressed with the almost endless variety of song it has to offer. The whirring noise the *'Apapane* makes with its wings also contributes to its repertoire. It is so distinctive that the species can sometimes be identified from that sound alone. Although it prefers the nectar of the *lehua* blossom, it can occasionally be seen in the subcanopy feeding on the native *kanawao*. The *'Apapane* is a quick-moving bird and not easily approached, probably because the native *Pueo* and introduced Barn Owl prey upon it. When an owl is in the vicinity, *'Apapane* hide within the terminal leaf clusters of the *'ōhi'a* and emit whimperlike calls. It can be seen throughout the Kōke'e area.

'Apapane, Himatione sanguinea sanguinea.

ʻIʻiwi

Vestiaria coccinea   5.5 inches (14 cm)

Other names given to this species are *ʻIʻiwi pōlena* and *ʻIʻiwi pōpolo*. On Kauaʻi the name *Olokele* was also used. Extinct on Lānaʻi, and virtually so on Molokaʻi and Oʻahu, the *ʻIʻiwi* is still common in the high native forest of Kauaʻi. An endemic honeycreeper, this species feeds among the curved blossoms of the endemic lobelias, like the *koliʻi*, a plant found only on Kauaʻi. Once valued for its bright red plumage, this handsome bird held an important role in Hawaiian culture. It is estimated that the feathers of 30,000 *ʻIʻiwi* were used to produce a single cape to adorn the *aliʻi*. A collector on a visit to Kauaʻi a century ago noted that native boys would catch the *ʻIʻiwi* by hiding in the bushes with a curved flower held between thumb and forefinger. When the bird inserted its long curved bill the boys would pinch and hold fast to their prize. The *ʻIʻiwi* can be seen at any time of year, but the best time to watch this species use its uniquely adapted bill is in the months of September and October, when it takes nectar at the *koliʻi* colony along the Pihea boardwalk.

ʻIʻiwi, Vestiaria coccinea.

Kāmaʻo

Myadestes myadestinus   8 inches (20 cm)
Ⓔ an endangered species

A hundred years ago, the *Kāmaʻo* or Large Kauaʻi Thrush was the most plentiful forest bird on the island. People from the seashore to the mountains enjoyed its melodious song. By the late 1920s it could no longer be found in the lower forests, and by the 1960s the population, which had shrunk to a few hundred birds, was restricted to the remote regions of the Alakaʻi. In the most recent forest bird survey conducted by state and federal biologists, the *Kāmaʻo* was not seen or heard. This thrush has not been reported since 1990 and may already be extinct. After thousands of years, the voice of the Large Kauaʻi Thrush has fallen silent. The only thrushes heard with regularity today in the forests of Kauaʻi are those introduced from foreign lands.

Kāmaʻo (Large Kauaʻi Thrush), *Myadestes myadestinus* [Illustration from Wilson and Evans, *Aves Hawaiiensis*].

Kaua'i 'Akialoa

Hemignathus ellisianus procerus   7 inches (18 cm)
Ⓔ an endangered species

In the 1890s collector George C. Munro (1944) stated that he found
this magnificent honeycreeper throughout the island, from the high
wet plateau to the forested seashore at Hanalei. A mere two decades
later he was unable to locate it. A few individuals were reported in the
1960s, but since then, despite exhaustive searches, the *'Akialoa* has not
been seen and is presumed extinct. The Kaua'i *'Akialoa* had, by far, the
most impressive bill of any historically known Hawaiian bird. Observers
noted that it used its remarkable bill to probe for insects in bark crevices
and to reach far down into the leaf bases of the native *hala pepe* and *'ie'ie*
plants. It was also reported to feed on the nectar of endemic lobelias.
Munro reported that many of the *'Akialoa* he saw at low elevations were
diseased. Some were so disabled with large sores on their bill and legs
that they could barely fly. Others had tumors in their throats or were
infested with tapeworms. European collectors learned that the male and
female *'Akialoa* formed strong pair bonds. When one bird was shot, the
other would repeatedly return to look for its mate until it too could be
collected.

Kaua'i *'Akialoa, Hemignathus
ellisianus procerus* [Illustration
from Wilson and Evans, *Aves
Hawaiiensis*].

Kaua'i '*Amakihi*

Hemignathus kauaiensis   4.5 inches (11 cm)

The Kaua'i '*Amakihi* is perhaps the most adaptable of Hawaiian honeycrecpers. With its multipurpose bill it feeds on the fruit of the introduced Java plum, seeks insects beneath the bark of trees, and takes nectar from the lobelias flowering in the dense undergrowth of the rain forest. Despite its apparent success, its range is still restricted. Like many other endemic birds, the '*Amakihi* cannot be found at low elevations. Each of the four large islands of Hawai'i has a race of this honeycreeper, but the curved bill of the Kaua'i species is the longest and most impressive. It can be recognized easily by its call—two distinct notes, the second higher in pitch. The '*Amakihi* is common in the Kōke'e area and easily seen along the Pihea trail.

'*Amakihi, Hemignathus kauaiensis.*

Kaua'i *'Elepaio*

Chasiempis sandwichensis sclateri   5.5 inches (14 cm)

The *'Elepaio* is a favorite of many bird-watchers because it shows little fear of humans. The bird, demonstrating its classic tail-up pose, will come within a few feet of an observer. An imitation of its call will often entice it to approach close enough to touch. Although the Kaua'i sub-species of the *'Elepaio* is less colorful than its O'ahu and Big Island cousins, it is no less active. It is an acrobatic flier that darts through the understory to catch insects in the air. As is common with many Old World flycatchers, the nest of the *'Elepaio* is spotted with lichens. The *'Elepaio* can be seen at elevations lower than any other native passerine; however, it is still restricted to the high forest. It can easily be found along the Pihea boardwalk and other Kōke'e trails.

Kaua'i *'Elepaio, Chasiempis sandwichensis sclateri.*

Kaua'i *Nukupu'u*

Hemignathus lucidus hanapepe   5.5 inches (14 cm)
Ⓔ an endangered species

This species was named for the fact that the bill *(nuku)* is shaped like a
hill *(pu'u)*. Historically, the *Nukupu'u* has never been very common. In
fact, the species was not discovered by Europeans until 1887. It is now
nearly extinct. Even a century ago, collector George C. Munro (1944)
reported that it was seldom seen below the wet Alaka'i plateau. In color
and size it so resembles other more common yellow-green honeycreepers that it could go unnoticed in the thick subcanopy of the wet forest.
In habits, it resembles the *'Akikiki* as it moves from tree to tree probing
mosses and bark crevices for mature insects and larvae. Unlike the
'Akikiki, however, the *Nukupu'u* reportedly taps on bark like a wood-
pecker. Observers wishing to report sightings of this species must be
careful, however, because it is easy to be misled by the thick, curved bill
and bark-peeling habits of the Kaua'i *'Amakihi*. The key to correctly
identifying this short-tailed honeycreeper is a good view of its long, thin,
curved bill. Some sources say the *Nukupu'u* rarely takes nectar, but a
bird seen by biologist Tom Telfer a decade ago in a remote valley of the
Alaka'i was feeding at the tubular flowers of the *hāhā lua*. This species,
if it still exists, numbers no more than a few individuals.

Kaua'i *Nukupu'u, Hemignathus
lucidus hanapepe* [Illustration
from Wilson and Evans, *Aves
Hawaiiensis*].

'Ō'ō 'ā'ā

Moho braccatus   8 inches (20 cm)
Ⓔ an endangered species

The Kaua'i *'Ō'ō* has been referred to as the "rarest bird in the world"
(National Geographic Video, *Hawai'i: Strangers in Paradise,* 1991). The
Hawaiian word *'ā'ā* means "dwarf" or "small" and was given to the Kaua'i
species because it was smaller than the three races found on the other
islands. It was thought that all of these birds became extinct near the
end of the nineteenth century. It was thus a surprise to many in 1960
when an isolated colony of the *'Ō'ō 'ā'ā* was found to still reside in the
remote Alaka'i. Biologist John Sincock studied this group of about two
dozen birds for many years. He banded the nesting trees to keep rats
from climbing them and even put up bird houses to provide more nesting cavities. Nothing seemed to work and he helplessly watched it
decline. The *'Ō'ō 'ā'ā* has not been seen since the late 1980s. The Kaua'i
species, the last of Hawai'i's magnificent *'Ō'ō,* has probably joined the
others of its genus on the growing list of birds that no longer exist.

'Ō'ō 'ā'ā, Moho braccatus
[Illustration from Wilson
and Evans, *Aves
Hawaiiensis*].

ʻŌʻū

Psittirostra psittacea   6.5 inches (17 cm)
Ⓔ an endangered species

According to Pukui and Elbert (1986) and Munro (1944), Hawaiians had two distinctive names for the male and female of this species: *ʻŌʻū poʻo lapalapa* (square-headed *ʻŌʻū*) and *ʻŌʻū lae oʻo* (mature-headed *ʻŌʻū*). Once plentiful, the bicolored *ʻŌʻū* is now critically endangered and not expected to survive into the twenty-first century. A shy bird that often perches high in the canopy, it has a loud call that it uses to communicate though the foggy mist that frequently envelopes the wet forest. Although there have been no confirmed sightings since 1987, loud whistles suggestive of the species have recently been heard by biologists, indicating that this bird may still cling to existence. The parrotlike bill of this honeycreeper allows it to feed on the fruit of the endemic lobelias as well as that of the *ʻieʻie* vine. It was also known to migrate to the lowlands to eat the fruit of the introduced strawberry guava, a habit that took it into mosquito-infested areas. The malaria and pox it acquired there may well have devastated the once great flocks of the *ʻŌʻū*.

ʻŌʻū, Psittirostra psittacea
[Illustration from Wilson and Evans, *Aves Hawaiiensis*].

Puaiohi

Myadestes palmeri   6.5 inches (17 cm)
Ⓔ an endangered species

Mr. H. C. Palmer, the collector for whom this bird is named, found the *Puaiohi* so rare in the forests of Kaua'i that he was able to attain only one specimen. R. C. L. Perkins (1903), another noted collector of the period, considered it one of the rarest birds he had ever pursued. Also known as the Small Kaua'i Thrush, the *Puaiohi* is still extremely rare and is listed as an endangered species. Like all Hawaiian thrushes, the *Puaiohi* is a bird of the subcanopy and forest floor, where it feeds on fruit and insects. It is one of the few native frugivores remaining in the Alaka'i and is thus an important link in the dispersion of endemic seeds in the ecosystem. As it was in the 1890s, the *Puaiohi* remains a very difficult bird to find. The species prefers deep, narrow valleys in the wet forest, where it nests in cavities in the steep, fern-covered valley walls. A cooperative effort between government agencies and a private conservation group, The Peregrine Fund, to propagate this species in captivity has given hope that the *Puaiohi* will increase and that it may one day be found with ease along Kaua'i's forest trails.

Puaiohi (Small Kaua'i Thrush), *Myadestes palmeri.*

Pueo

Asio flammeus sandwichensis 13–17 inches (33–43 cm)

A subspecies of the Short-eared Owl, the *Pueo* or Hawaiian Owl was regarded with reverence by the ancient Hawaiians. It is interesting that fossil bones of this species do not appear in Hawaiian deposits until after Polynesian arrival. Scientists theorize that earlier attempts at colonization may have failed because the open grassland habitat favored by this species was probably not present until after humans altered the Hawaiian landscape. The *Pueo* hunts around daybreak, at sunset, and sometimes at midday. Although it frequents lowland fields and pastures, this endemic species can be found in the native forest. Its diet is mostly rodents and insects, but the *Pueo* does prey upon both native and alien birds. When hunting it hovers above its prey before diving down to attack. The *Pueo* can be seen in most rural open areas and occasionally in towns at dusk. Look for them near Kīlauea Point, from Kaumakani to Mānā, and along the lower Kōke'e road. Deep, penetrating yellow eyes and a round facial disk distinguish this species from the Barn Owl.

Pueo (Hawaiian Owl), *Asio flammeus sandwichensis.*

Wetland Birds and Shore Birds

EVERY SPECIES of native wetland bird that was here in 1778 can still be found in the taro patches, streams, and wildlife refuges around the Garden Island. This can be said of no other Hawaiian island except Oʻahu and is due, in part, to the absence of the mongoose on Kauaʻi. It has been said that when mongooses were shipped to Kauaʻi to combat rats in the sugarcane fields, a worker on the docks was inadvertently bitten by one of the furry creatures and angrily threw the crate overboard. True or not, Kauaʻi is fortunate not to have this destructive mammal, which eats the eggs and young of ground-nesting birds. Nevertheless, many of the birds included in this section are listed as endangered species. Feral cats, dogs, and rats continue to prey upon them, and much of Kauaʻi's once extensive lowland marsh has succumbed to agriculture and development. The good news is that a century-long trend of shrinking habitat is slowly being reversed. Through the creation of sanctuaries and refuges the State Department of Land and Natural Resources and the U.S. Fish and Wildlife Service have actually increased the amount of wetland habitat available to these birds.

In addition to the six resident native species, more than twenty species of ducks, geese, and shorebirds migrate to Hawaiʻi each winter. Scientists speculate that it was from seeds attached to the legs of these visitors from far-off lands that Hawaiʻi received many of the colonizing plants that now make up the lush, native rain forest. More than that, it has been suggested that the *Kōlea* or Pacific Golden-Plover may have been responsible for the discovery of Hawaiʻi. Polynesians may have sailed north from the distant Marquesas and Society Islands, where the *Kōlea* also winters, to search for the place from whence the bird came.

Good places to find these birds are on the ponds and taro patches of the Hanalei National Wildlife Refuge, at the Wailua Reservoir, at Hanapēpē Salt Pond (after heavy rains), at Kawaiele Bird Sanctuary

The Hanalei National Wildlife Refuge has proven to be a successful cooperative venture between the government and the taro farmers who use it. Wetland species like the *Ae'o* (Hawaiian Stilt) do well in the cultivated fields that cover the valley floor.

near Mānā, at Kaua'i Lagoons near Nāwiliwili, and at Smith's Tropical Paradise on the Wailua River. Waitā Reservoir, a location recommended in several bird guides, has recently been posted with No Trespassing signs and is no longer open to the public.

Ae'o

Hawaiian Stilt

Himantopus mexicanus knudseni   16 inches (41 cm)

Ⓔ an endangered species

The *Ae'o* is a subspecies of the Black-necked Stilt found in North America. The endemic Hawaiian race has more black on the face and neck than the mainland form. The *Ae'o* is a strong flier and is known to migrate to and from Ni'ihau. In winter months they can often be seen feeding singly or in small groups in fields flooded by heavy rains. It defends its nesting area vigorously and will sometimes feign injury to draw attention away from the nest. According to *Hawaii's Birds*, a publication of the Hawai'i Audubon Society, only 1,800 birds remain.

Ae'o (Hawaiian Stilt), *Himantopus mexicanus knudseni.*

'Akekeke

Ruddy Turnstone

Arenaria interpres   9.5 inches (24 cm)

Red legs and a black bib make it easy to distinguish this robust shore bird from Kaua'i's other winter migrants. After a summer of nesting in the Arctic tundra, the Ruddy Turnstone spends the months of August through April in Hawai'i and on coastal areas throughout the tropical Pacific. Unlike the solitary *Kōlea,* the *'Akekeke* prefers to feed in pairs or small groups. At times these groups can be quite large. A flock of turnstones seen at Salt Pond in Hanapēpē numbered fifty birds. Ruddy Turnstones, as their name suggests, feed by flipping over rocks with their bill to find the invertebrates that lie beneath. It is rarely seen on sandy beaches, preferring instead pebble-strewn beaches, mudflats, or flooded pastures and grasslands.

'Akekeke (Ruddy Turnstone), *Arenaria interpres.*

'Alae ke'oke'o

Hawaiian Coot

Fulica alai   14 inches (36 cm)

Ⓔ an endangered species

The endemic Hawaiian Coot is descended from the American Coot, a species that breeds from southern Canada to Panama and that occasionally migrates to the Hawaiian Islands. The endemic *'Alae ke'oke'o* is named for its white *(ke'oke'o)* frontal shield. It differs from the ancestral American Coot and other coots of the world in the distribution of white on the head. The Hawaiian Coot, like the Hawaiian Stilt, is known to fly to Ni'ihau and even nests there when sufficient water is present. Hunters in the 1800s remarked that the *'Alae ke'oke'o* could be found on Kaua'i by the "millions." Its numbers are now estimated at only a few thousand.

'Alae ke'oke'o (Hawaiian Coot), *Fulica alai.*

'Alae 'ula

Hawaiian Gallinule

Gallinula chloropus sandvicensis   13 inches (33 cm)
Ⓔ an endangered species

Also known as the Common Moorhen or Mudhen, the *'Alae 'ula* gets its
Hawaiian name from its red *('ula)* frontal shield. The endemic Hawaiian
subspecies is descended from the Common Moorhen of North America.
The *'Alae 'ula* is a shy bird that prefers marshes or ponds with thick mar-
ginal vegetation. It can no longer be found on the Big Island or Maui
and attempts to reintroduce it there have failed. The secretive behavior
of this bird makes it difficult to enumerate, but the *'Alae 'ula* on Kaua'i
is thought to number in the low hundreds.

'Alae 'ula (Hawaiian Gallinule), *Gallinula chloropus sandvicensis.*

'Auku'u

Black-crowned Night-Heron

Nycticorax nycticorax hoactli   25 inches (63 cm)

Indigenous to the Hawaiian Islands, the *'Auku'u* is a solitary wading bird that often stands motionless on the banks of ponds and canefield ditches while eyeing its prey in the water below. The *'Auku'u* is known to feed in both daylight and darkness. When flying they sometimes utter a loud "kwok" noise. Immature birds are heavily streaked with brown and white. Breeding adults are adorned with a long, white head plume. The *'Auku'u* arrived in the Hawaiian Islands naturally, has bred here for centuries, and is thus considered a native bird; however, the Hawaiian form is as yet indistinguishable from the North American species.

Left: *'Auku'u* (Black-crowned Night-Heron), *Nycticorax nycticorax hoactli;* adult. Right: *'Auku'u;* juvenile.

Hunakai

Sanderling

Calidris alba   8 inches (20 cm)

The Sanderling is a joy to watch when seen on one of Kaua'i's many sandy beaches. White and small, it runs back and forth at the water's edge, actively probing with its bill for invertebrates hidden in the wet sand. The name *Hunakai* means "sea foam," an appropriate name for this perpetually moving shore bird. It is so adept at staying just ahead of the waves that it almost seems to be a part of the foam at the leading edge. Its little black legs move so fast they are a blur. Some have described these movements as being like a wind-up toy. The *Hunakai* feeds both as an individual and in small flocks of three or four birds. After heavy rains, Sanderlings can be found even in flooded pastures. This winter visitor migrates throughout much of the tropical Pacific. It winters on Kaua'i from August through April.

Hunakai (Sanderling), *Calidris alba.*

Kōlea

Pacific Golden-Plover

Pluvialis fulva   11 inches (28 cm)

Once called the Lesser Golden-Plover, the *Kōlea* is the most common winter visitor on Kaua'i. This species arrives in Hawai'i in August after spending May, June, and July nesting in the Arctic. It is a common sight on golf courses, pastures, and other open spaces. The *Kōlea* can be seen even in the bogs of the wet Alaka'i. The long, curving, white stripe bordered by black is breeding plumage, which is most prominent in April. Studies on banded individuals have shown that the *Kōlea* returns to the same territory year after year. In the winter months they are often seen as solitary birds, but as the time nears for them to depart, they gather into large flocks in preparation for the trip north.

Kōlea (Pacific Golden-Plover), *Pluvialis fulva:* nonbreeding plumage.

Kōlea: breeding plumage.

Koloa maoli

Hawaiian Duck

Anas wyvilliana   Male: 20 inches (51 cm); female: 17 inches (43 cm)
(E) an endangered species

This endemic waterfowl nests along the banks of mountain streams, taro patches, estuaries, and isolated irrigation ditches. The Hawaiian Duck is closely related to the migratory Mallard and resembles the female of that species. There is very little difference in the color pattern between male and female *Koloa maoli;* however, males are somewhat larger and have a darker head. In both sexes, the legs and feet are orange. In the 1840s this duck was so plentiful that it was slaughtered by the hundreds to provision visiting whaling ships. Very leery of humans, the *Koloa maoli* is shy and difficult to observe. Approachable individuals of this species occasionally frequent Kaua'i Lagoons near Nāwiliwili and the taro patches in Hanalei. On other islands, *Koloa*/Mallard hybrids do exist. Kaua'i remains the only place in Hawai'i where genetically pure *Koloa* can still be seen easily.

Koloa maoli (Hawaiian Duck), *Anas wyvilliana.*

Koloa māpu

Northern Pintail

Anas acuta   26 inches (66 cm)

Like the Northern Shoveler, this handsome waterfowl is one of the more common migrants to Kaua'i. Males are easily recognized by the prominent white stripe that extends up the neck and onto the head. Both sexes have pointed pinlike tails. The Northern Pintail once migrated to Hawai'i by the thousands. Observers in the 1800s noted that there were large flocks of them on the lagoons and ponds of Kaua'i. One historian interprets the Hawaiian word *māpu* as "to rise or float off as a cloud," a description that adequately describes hundreds of these ducks rising off the water to take flight. Other linguists prefer to interpret *māpu* as "wind-blown" or "swooping," also a good description of this species, which often appears on Kaua'i after strong winds.

Koloa māpu (Northern Pintail), *Anas acuta*.

Koloa mohā

Northern Shoveler

Anas clypeata   19 inches (48 cm)

Many species of ducks migrate to Hawai'i, but the Northern Shoveler winters in the Islands with such regularity that it has been given a Hawaiian name. *Mohā* (bright) describes the glossy green head of males in breeding plumage. The best way, however, to identify the species is not by color, but by the shape and length of the bill. It is the only duck in which the bill, shaped like a wide spoon or a shovel, is longer than the head. The Northern Shoveler is a member of a group of ducks known as "puddle" or "dabbling" ducks. All the members of this group prefer shallow freshwater or brackish ponds, where they tip over, tail up, to grub the bottom for food.

Koloa mohā (Northern Shoveler), *Anas clypeata*.

Long-billed Dowitcher

Limnodromus scolopaceus   11.5 inches (29 cm)

This winter migrant is an exciting find for Hawai‘i birders. The impressive long bill of this dowitcher easily sets it apart from other more common visitors like the Pacific Golden-Plover, Ruddy Turnstone, and Sanderling. It may be confused with the *‘Ūlili* (Wandering Tattler), which also has a long bill and yellow legs, but the feeding habits of the two species are very different. The Long-billed Dowitcher feeds by probing with its bill in a manner that has been described as like the rapid up-and-down motion of a sewing machine needle. Look for it on mudflats and along the shorelines of freshwater ponds and taro patches. A similar species, the Short-billed Dowitcher, whose bill length overlaps that of the Long-billed Dowitcher, has never been documented on Kaua‘i. The plumage of the two species is virtually indistinguishable in winter months. To accurately resolve the two species, experts recommend that you listen to the call of the dowitcher when disturbed. The long-billed species gives a "keeek" call, whereas the call of the Short-billed Dowitcher has been described as "tu-tu-tu."

Long-billed Dowitcher, *Limnodromus scolopaceus.*

Nēnē

Hawaiian Goose

Branta sandvicensis   24 inches (61 cm)
(E) an endangered species

By the time Captain Cook arrived on Kaua'i, this endemic goose did not exist here; however, fossil records gathered by Smithsonian researchers Storrs Olson and Helen James have confirmed that the *Nēnē* or Hawaiian Goose once thrived in Kaua'i's lowlands. It is presumed to have been extirpated by humans before European arrival. It was reintroduced a few decades ago and is doing well. In fact, scientists believe that the reintroduced Kaua'i population is now the most robust and the least likely to go extinct. Flocks can be seen in the Kīlauea area, in the Kōke'e meadow, and at Kaua'i Lagoons in Nāwiliwili. Although this species is listed here as a native wetland bird, the *Nēnē* spends little time in the water, preferring instead open grassy areas, pastures, and golf courses, where it feeds on a variety of native and introduced plants.

Nēnē (Hawaiian Goose),
Branta sandvicensis.

'Ūlili

Wandering Tattler

Heteroscelus incanus   11 inches (28 cm)

The Wandering Tattler, as its English name suggests, has an extensive migratory range that includes most of the tropical Pacific. It arrives on Kaua'i in late August and remains until April or May. The summer months are spent nesting in Alaska or Canada. Unlike the *Kōlea*, this winter visitor is rarely seen in grassy fields, preferring instead mudflats and, especially, coastal areas, where it probes in the crevices of wave-washed rocks with its long, thin bill. When the *'Ūlili* is disturbed it utters its name on takeoff—"u-li-li-li." The popular Hawaiian song "'Ūlili E" (author unknown) beautifully describes the voice and actions of this regular visitor to Kaua'i (English translation by Hawaiian language instructor Paul Williams):

Hone ana ka leo e 'Ūlili e   The voice of the *Ūlili* is sweet,
'Ekahi manu noho 'ae kai   A bird that stays by the water's
                                          edge.

Kia'i ma ka lae a'o Kekaha   Guarding the point at Kekaha
'O ia kai ua lana mālie.   This gently flowing sea.

'Ūlili (Wandering Tattler), *Heteroscelus incanus.*

Other Migratory Species Reported Regularly from the Shores and Wetlands of Kaua'i

American Wigeon, *Anas americana*
Green-winged Teal, *Anas crecca*
Kioea (Bristle-thighed Curlew), *Numenius tahitiensis*
Lesser Scaup, *Aythya affinis*
Mallard, *Anas platyrhynchos*
Pectoral Sandpiper, *Calidris melanotos*
Ring-necked Duck, *Aythya collaris*

Occasionally a surprise will show up as with the Great Blue Heron that visited the salt ponds near Hanapēpē several years ago. Few of these migratory birds are approachable, so a spotting scope or binoculars are a must when trying to decide what it is that you see. The *Field Guide to the Birds of North America* published by the National Geographic Society is a good companion to take along.

Kawaiele Bird Sanctuary, Mānā. A project of the State Department of Land and Natural Resources, this wetland is home to several species of endangered wetland birds.

Seabirds

SEABIRDS ARE not only inspiring to watch as they gracefully glide on the air currents above the ocean surface, but they are a valuable asset to local fishermen, who depend on "bird piles" to locate the large *'Ahi* (Yellow-fin Tuna) that frequent the waters around Kaua'i. "Bird piles" are large gatherings of mixed-species flocks that feed on the same fish pursued by the *'Ahi*.

Short of being in the middle of one of these congregations there is no better place in the Hawaiian Islands to view them up close than the Kīlauea Point National Wildlife Refuge. Of the twenty species of seabirds that are known to frequent the vicinity of Kaua'i, Ni'ihau, Lehua, Ka'ula, and Moku'ae'ae (a small island directly off Kīlauea Point), six regularly can be seen at this scenic peninsula, and two others (the White-tailed Tropicbird and the Brown Booby) occasionally show up.

Nine species of seabirds currently breed on Kaua'i and are thus considered native to this island. In ancient times all twenty, or even more, may have nested here. That these nesting birds were an easy source of food for the ancient Hawaiians is a fact borne out by the thousands of bones found in midden deposits around ancient house sites. The fledgling *'Ua'u* (Dark-rumped Petrel) was considered such a delicacy that it was reserved for the *ali'i* (ruling chiefs). Today, predation by rats, cats, dogs, and pigs is the principal cause of small numbers of nesting seabird colonies on Kaua'i.

Nevertheless, there is reason for optimism. The *Mōlī* (Laysan Albatross) has returned to Kaua'i and is even known to nest at the Pacific Missile Range Facility at Barking Sands. The purchase of land adjacent to the Kīlauea lighthouse and the fencing off of this unique refuge has substantially increased breeding space for this species and has reduced predation by feral animals.

Another success story concerns the endangered *'A'o* (Newell Shearwater). Each fall, shearwaters, disoriented by the lights of Kaua'i's

The Kīlauea National Wildlife Refuge is the only place in the state where an observer can come within a few feet of graceful, soaring seabirds.

urban areas, fall onto the island's highways by the hundreds. In 1978, the State Department of Land and Natural Resources instituted an effort to recover these young birds and set them aloft again. More than 25,000 birds have been picked up off the roads by concerned travelers, placed in cages provided at the island's fire stations, banded, and released by personnel of the DLNR.

The Kīlauea Point National Wildlife Refuge is open to the public every day including Saturday and Sunday. Hours of operation are 10 A.M. to 4 P.M. The entrance fee of $2 per adult is collected in a drop box via the honor system. Volunteer docents are available at the Lighthouse Visitor Center to answer your questions. A docent-led, interpretive hike is also available. Call 808–828–0168 to inquire about this informative hike.

Each fire station in the County of Kaua'i has a rack of cages in which residents can deposit fallen seabirds.

ʻĀ

Brown Booby

Sula leucogaster plotus   28 inches (71 cm)

The Hawaiians apparently did not distinguish the three species of boobies that frequent the coastal waters of Hawaiʻi because they gave them all the same name, ʻĀ. Although the Brown Booby does not nest on Kauaʻi, it is known to roost on Mokuʻaeʻae, a small island off Kīlauea Point. Red-footed and Masked Boobies (the third species [*Sula dactylatra personata*] and a rare visitor to Kauaʻi's coast) are pelagic feeders and often mix with other seabirds. The Brown Booby does not mix with other species and usually feeds near the shore. It can frequently be seen plunge diving off the leeward coast from Waimea to Polihale.

ʻĀ (Brown Booby), *Sula leucogaster plotus*.

'Ā

Red-footed Booby

Sula sula rubripes   28 inches (71 cm)

The Hawaiian name for the indigenous Red-footed Booby is a mimic of its monotonous cry. It gets its unflattering English name from sailors who believed that these birds were stupid because they were so easily grabbed by the feet when landing on the ship's rigging. The 'Ā is designed for diving beneath the surface of the sea to capture its prey and will plunge into the ocean from heights of up to a hundred feet (30 m) to begin its pursuit. In spring and summer months, the Crater Hill slope adjacent to Kīlauea Point is littered white with the presence of hundreds of these nesting birds.

'Ā (Red-footed Booby), *Sula sula rubripes.*

'Akē'akē

Band-rumped Storm-Petrel

Oceanodroma castro   8 inches (20 cm)

The indigenous Band-rumped Storm-Petrel is the smallest seabird in Hawai'i. Also called the Harcourt Storm-Petrel, it frequents both the Pacific and Atlantic Oceans, but is so rare in Hawaiian waters that its nest and breeding behavior here have never been described. Kaua'i and Maui are the only places in Hawai'i that it is known to nest, although birds have been heard at night at high altitudes on Mauna Loa on the Big Island. One or two are found each year on Kaua'i's highways. Storm-petrels are so named for their habit of hiding in the lee of ships during storms at sea. Sailors considered them an ill omen and thought they were gathering to carry off the souls of the drowned. Storm-petrels do not dive into the water. They prefer instead to flutter near the surface when feeding. Rats are thought to be a serious predator of this small seabird.

'Akē'akē (Band-rumped Storm-Petrel), *Oceanodroma castro*.

'A'o

Newell Shearwater

Puffinus newelli   13 inches (33 cm)
Ⓔ an endangered species

To the Hawaiians who named it, the moan of the 'A'o in its burrow
sounded like an omen of death. The endemic Newell Shearwater has
become so rare that few can attest to ever having heard its nesting cry.
The 'A'o nests high in the mountains from April to November, where it
conceals its burrow beneath the thick *uluhe* fern. When fledglings
depart for a life at sea, many become confused by the lights of developed areas and fall upon Kaua'i's highways. Although Kaua'i residents
salvage many of these birds each year in a highly acclaimed recovery
program, the number of 'A'o continues to decline. Feral cats, which kill
the nesting birds in their burrows, have a devastating impact on this
species.

'A'o (Newell Shearwater), *Puffinus newelli.*

'Iwa

Great Frigatebird

Fregata minor palmerstoni   43 inches (109 cm)

The *'Iwa* is often described as an unparalleled flying machine for its ability to outmaneuver other seabirds. Weighing only 3 pounds (1.4 kg) and with a 7-foot (2.1 m) wing span, this forked-tailed giant can easily soar as high as 500 feet (150 m)—higher than any other seabird. The Hawaiian word *'iwa* means thief. Great Frigatebirds are adept at stealing food from Red-footed Boobies and shearwaters as they return with food to the nest. From great heights, they swoop down to harass their victims until they disgorge their meal. Sailors of old referred to them as "man-o'-war birds" because they reminded them of the swift frigate pursuit ships used by pirates in the tropical seas. Great Frigatebirds travel widely. Birds banded in Hawai'i have turned up in the Philippines. Although commonly seen above the island's coastal waters and occasionally resting on Moku'ae'ae, the *'Iwa* does not nest on Kaua'i.

'Iwa (Great Frigatebird), *Fregata minor palmerstoni.*

Koa'e kea

White-tailed Tropicbird

Phaethon lepturus dorotheae   27 inches (69 cm)

The indigenous *Koa'e kea* or White-tailed Tropicbird nests on Kaua'i throughout the year. It prefers to build its nest, which is actually not much more than a scrape, in cavities in steep cliff faces. Although they can occasionally be seen at the Kīlauea Lighthouse, they are more often observed circling within Waimea Canyon or high against the verdant cliffs of the precipitous north shore *pali*. At great distances when tail and bill color cannot be discerned, the *Koa'e kea* can be distinguished easily from its red-tailed cousin: The white-tailed species has black wing bars when viewed from above.

Koa'e kea (White-tailed Tropicbird), *Phaethon lepturus dorotheae.*

Koaʻe ʻula

Red-tailed Tropicbird

Phaethon rubricauda rothschildi   39 inches (99 cm)

Named for its long red tail feathers, the indigenous *Koaʻe ʻula* or Red-tailed Tropicbird is a beautiful sight to watch during the mating season of March to October. In an elaborate courtship display, it flies backwards, as if rowing a boat, and at times the bird is nearly stationary on the wind. The Red-tailed Tropicbird nests under beach vegetation, and although graceful in the air, the *Koaʻe ʻula* is so awkward on land that it cannot stand up on its own feet or walk forward without falling. Like boobies, tropicbirds dive headlong into the water from great heights and are known to go as deep as 10 feet (3 m) to catch fish. The *Koaʻe ʻula* is rare on the other main islands. It can only be seen with ease at the Kīlauea National Wildlife Refuge.

Koaʻe ʻula (Red-tailed Tropicbird), *Phaethon rubricauda rothschildi.*

Laughing Gull

Larus atricilla   16.5 inches (42 cm)

Most sightings of seagulls in Hawai'i occur during the winter months, a time when these stragglers lack the characteristic plumage of their normal confines. The Laughing Gull is an excellent example. In summer months, on the eastern coast of North and Central America where this species breeds, the adult Laughing Gull has a distinctive black hood and dark red bill. Birds seen in Hawai'i are rather drab. This species closely resembles the Franklin Gull *(Larus pipixcan)*, a much rarer visitor to Hawai'i. To accurately identify any gull seen on Kaua'i, pay close attention to details of bill color and markings, leg color, and plumage in flight. *Gulls: A Guide to Identification* by P. J. Grant (1997) is an excellent text to consult when trying to sort out these occasional visitors to Kaua'i.

Laughing Gull, *Larus atricilla.*

Mōlī

Laysan Albatross

Phoebastria immutabilis 32 inches (81 cm)

More commonly called the Laysan Albatross, the *Mōlī*, with a wing span of 80 inches (2 m), is unmistakable in the air as it effortlessly glides on the winds near the surface of the ocean. Its clumsiness on land, however, has led some to call it the gooney bird. The *Mōlī* is making a comeback on Kaua'i. Birds are not only nesting at Kīlauea Point but also on the grounds of the Pacific Missile Range Facility on the island's west side. After spending the summer months over a thousand miles (1,600 km) out at sea in the cold waters of the North Pacific, it is a mystery how this magnificent indigenous bird can return to nest, not only to the same island, but to the exact spot from which it hatched. Each seabird in Hawai'i has a different method of feeding. Albatrosses are not divers—the bird lands on the water and fishes from the sitting position. They are sometimes attacked from below by large fish and sharks.

Left: *Mōlī* (Laysan Albatross), *Phoebastria immutabilis*. Right: *Mōlī* in flight.

Ring-billed Gull

Larus delawarensis   17.5 inches (44 cm)

Sea gulls are a common sight along continental coastlines, but for reasons not well understood, none of the world's nearly fifty species of gulls has become established in Hawai'i. In fact, since records have been kept, only thirteen species have ever been sighted in the Hawaiian Islands. Gulls prefer broad, shallow tidal zones, conditions not found on tropical islands. One of the few that does occasionally visit is the Ring-billed Gull. Plumage varies with the age of the bird, but the ring near the tip of the bill is present in all but juveniles. An excellent place to look for it and others is at the Kawaiele Bird Sanctuary near Mānā on Kaua'i's west side.

Ring-billed Gull, *Larus delawarensis*.

'Ua'u kani

Wedge-tailed Shearwater

Puffinus pacificus chlororhynchus   17 inches (43 cm)

The indigenous *'Ua'u kani* is named for the sound emitted from its burrow when nesting—"uaauuu"—a long, ghostly moan. English sailors named it the Wedge-tailed Shearwater because it seemed to them to glide so close to the water that the surface was cut or sliced. Hawaiian Wedge-tailed Shearwaters may spend as many as four years at sea, venturing as far as Panama or Japan. Like the Laysan Albatross, when it is time to breed, the *'Ua'u kani* has the uncanny ability to find its way back to the site where it hatched.

'Ua'u kani (Wedge-tailed Shearwater), *Puffinus pacificus chlororhynchus.*

Other Established Species Occasionally Seen from Shore

'Ua'u (Dark-rumped or Hawaiian Petrel), *Pterodroma phaeopygia sandwichensis.* This endemic subspecies breeds in Hawai'i. A closely related species breeds in the Galápagos. Each population is considered endangered. The only confirmed nesting sites in Hawai'i are on Maui near the summit of Haleakalā, at Mauna Loa on the Big Island, and on Kaua'i's Wainiha *pali.* The extent of the Kaua'i population is unknown. Each year during nesting season, a few of the precious birds are found on Kaua'i's highways.

Noio (Black Noddy), *Anous minutus melanogenys.* The most likely area to see the *Noio* near shore is at the end of the road at Hā'ena. This indigenous species occasionally is seen flying with quick wing beats searching for fish near the shoreline. Like storm-petrels, noddies dabble near the surface when fishing, barely getting their feet wet. The *Noio* makes its nest in cavities in the precipitous Nāpali coast, where the number of breeding pairs is estimated at 250 to 500 birds

Sitting directly off the end of Kīlauea Point, the island known as Moku'ae'ae is also a refuge for seabirds. The Brown Booby is known to roost here.

Other Nonbreeding Species Occasionally Seen from Shore

Black-footed Albatross, *Phoebastria nigripes*
Bonaparte Gull, *Larus philadelphia*
Bulwer Petrel, *Bulweria bulwerii*
Christmas Shearwater, *Puffinus nativitatis*
'Ewa'ewa (Sooty Tern), *Sterna fuscata oahuensis*
Glaucous-winged Gull, *Larus glaucescens*
Noio kōhā (Brown Noddy), *Anous stolidus pileatus*
Pākalakala (Gray-backed Tern), *Sterna lunata*

The Birds of Hawaii and the Tropical Pacific by H. Douglas Pratt, Phillip L. Bruner, and Delwyn G. Berrett (1987) is an excellent reference for identifying the many seabirds that can be seen in the Hawaiian Archipelago.

Introduced Birds

THE FIRST alien bird to be introduced intentionally to Kaua'i came over a millennium ago via the colonizing Polynesians. Along with their life-sustaining cargo of dogs and pigs was the *Moa* or Red Junglefowl. It is still doing well in the uplands of Kaua'i. With the second "discovery" of Kaua'i in 1778, the flood gate of introductions opened wide. The multitude of foreign ships that arrived in the wake of Captain Cook brought with them the animals and plants of their homeland. Among the pigeons and domestic chickens were other creatures heretofore unknown in the Hawaiian Islands—grazing mammals. Having evolved in the absence of herbivores, Hawaiian plants had no thorns or poisons to deter the newly arrived cows and goats that feasted upon them. As native plants were eaten, more aggressive alien species took root. Within decades, the native forest began to recede, forcing endemic forest birds into upland habitats. The lowlands were left so devoid of birds that people missed their joyful sounds. In an August 1860 article in the Honolulu *Pacific Commercial Advertiser,* a plea went out: "Owners of vessels leaving foreign ports for Honolulu, will confer a great favor by sending out birds, when it can be done without great expense. We need more songsters here."

After the population crash of Hawaiian birds at the turn of the century, an organized movement took form to "rehabilitate" the birds of the Hawaiian territory. The following article was published in the 19 March 1930 edition of *The Honolulu Advertiser:* "The practicability of the rehabilitation of bird life in Hawaii . . . has already been demonstrated by Mrs. Dora Isenberg on the Island of Kauai. In recognition of her efforts along this line, Mrs. Isenberg has been made a vice-president of Hui Manu, the avian society. Mrs. Isenberg has been interested for many years in bringing birds back to the island on which she lives, and as a result, the valleys near Lihue are filled with songsters."

"Rehabilitation" meant introducing new species to replace the old.

Lowland stream valleys like Waimea Canyon provide good habitat for introduced birds. At least thirty-three alien species now reside in Kaua'i's lowlands.

The efforts of the Hui Manu were to have a tremendous impact on bird populations in Honolulu and on the outer islands. In addition to the thirty-three species of alien birds now known to breed on Kaua'i, there were many other releases that did not take well to the Garden Island. Among them were tinamous from Chile, prairie chickens from North America, quails from Australia, pigeons from the Philippines, and larks from Mongolia.

Today, the State Department of Agriculture has strict regulations regarding the importation of any alien species. Even if a permit is obtained to bring in a bird from the "conditionally approved" list, that imported bird must remain in captivity. These laws are necessary to protect endemic birds. There are those who would love to make Kaua'i more of a paradise by having a colorful parrot in every tree, but it is important to remember that each released or escaped bird carries with it the potential of bringing a new disease or parasite to afflict dwindling native species or new competitors for food and space in the Alaka'i.

Barn Owl

Tyto alba   14–20 inches (36–51 cm)

Known also as the Common Barn Owl, this species was first released on Kaua'i at Kilohana in June 1959, and then in 1963 on lands owned by the Kekaha Sugar Company. The Barn Owl was imported in an effort to control rats in the sugar fields. Rodents do make up a large part of its diet; however, the Barn Owl also preys on birds and has been responsible for nesting failures in the Alaka'i and at Kīlauea Point. Although it can occasionally be seen hunting during daylight hours, this species is mostly a night predator. Sexes differ in size and in the amount of white in the face, but both male and female of this species have dark eyes and a heart-shaped facial disk. The only other owl present on Kaua'i is the native *Pueo,* which has yellow eyes in a round, dark face.

Left: Barn Owl, *Tyto alba;* male. Right: Barn Owl; female.

Black Francolin

Francolinus francolinus   13 inches (33 cm)

Introduced to Hawai'i in 1959, the Black Francolin has increased remarkably on Kaua'i in the last few years. The species favors dry habitats, and the voice of this bird is now a common sound in westside fields and dry scrub. The species has also been reported in Anahola. The early morning and late afternoon hours are the best time to hear it. The loud, buzzy, high-pitched call carries a long distance and is unlike that of any other bird on Kaua'i. The handsome black male has a large, white cheek patch and prefers to sing from elevated perches. The Black Francolin is wary and difficult to approach. This chickenlike bird can occasionally be seen flying across the highway.

Black Francolin, *Francolinus francolinus.*

Cattle Egret

Bubulcus ibis   20 inches (51 cm)

State officials imported the Cattle Egret from Florida around 1959 to help control insect pests on Island cattle ranches. Local children call it the "rubbish dump bird" because it congregates there by the hundreds. It frequently follows machinery in parks and fields to ingest the insects that are stirred up. Cattle Egrets are known to eat almost anything and pose a threat to the nestlings of waterbirds and seabirds. Fortunately, it is rarely found in the upland forest. As they fly to their communal roosting sites late in the day, flocks of these white birds form beautiful scenes against the lush greenery of Kaua'i's mountains.

Cattle Egret, *Bubulcus ibis.*

Chestnut Mannikin

Lonchura malacca   4.5 inches (11 cm)

The Chestnut Mannikin is an escaped cage bird that was first recorded on Kaua'i in 1975. Also known as the Black-headed Munia, this species has increased dramatically since its introduction and is now a common sight in the tall, green grass along roadsides and canefield ditches. At this writing, this species may well be Kaua'i's most plentiful lowland bird. The bird feeds in flocks that resemble a cloud of bees. A Chestnut Mannikin will land on the tall seed-bearing stalk of a weed to bend it over with its weight. It then walks out to the end to eat the seed. Mannikins frequently associate with Java Sparrows. The immature of this species is light brown with a blue-gray bill.

Chestnut Mannikin, *Lonchura malacca.*

Chukar

Alectoris chukar 13 inches (33 cm)

Imported from Asia in 1923, this shy game bird is rarely seen. It is currently established only in a few arid, rocky valleys on the leeward side of Kaua'i. It has possibly become rare in recent years because of interspecific competition with the Erckel Francolin.

Chukar, *Alectoris chukar.*

Common Myna

Acridotheres tristis   9 inches (23 cm)

This species has been here for generations and is Kaua'i's most easily recognized bird. Introduced in 1865 from India in an attempt to control army worms in pasture lands, the myna is now common throughout the state. It is a very social bird, traveling in pairs or small groups. Hundreds noisily gather to roost at dusk in large banyan or monkeypod trees. The myna does not hop like cardinals or finches. It prefers to walk or skip instead. Occasionally, they will gather in parks by the dozens for posturing exhibitions or what local children describe as "myna bird fights." Mynas harbor heavy infestations of bird mites that sometimes infest humans near their nest sites. It is common in lowland and upland habitats but has not penetrated deeply into the wet forest.

Common Myna, *Acridotheres tristis*.

Domestic Chicken

Gallus domesticus   Male: 30 inches (76 cm); female: 17 inches (43 cm)

It is not known when the first Domestic Chicken was introduced to Kaua'i. Sailors often took them on board ships as a source of food, so they could have been one of the first introductions after European "discovery." They are common along Kaua'i's roadways and come in a broad range of colors. In the aftermath of Hurricane 'Iniki (1992), which destroyed the cages of many captive birds, the Domestic Chicken has established breeding populations in the wild and their presence has increased dramatically.

Domestic Chicken, *Gallus domesticus.*

Erckel Francolin

Francolinus erckelii   16 inches (41 cm)

Some local people call this species the Laughing Bird. When it calls from concealment it gives the impression that someone is having a good guffaw at your expense. Introduced from Africa to Hawai'i in 1957, the Erckel Francolin is a favorite game bird of Kaua'i hunters. It favors high grasses in dry upland habitats but can be found in the wet forest as well. Francolins like to walk out onto roads in the early morning and late afternoon. A good place to see them is on the lower Kōke'e road an hour or so after sunrise.

Erckel Francolin, *Francolinus erckelii.*

House Finch

Carpodacus mexicanus   5.5 inches (14 cm)

Often called the Linnet or Papaya Bird (because of its fondness for overripe papaya fruit), the House Finch is one of the most common birds on Kaua'i. It is seen in large numbers in towns and agricultural areas. It can also occasionally be seen in small family groups within the native forest. Male House Finches on Kaua'i have predominately orange-red coloration on the breast and head; however, yellow-tinged birds have been seen. Female House Finches resemble female House Sparrows, but female finches are more heavily streaked. This species was imported from North America about 1869.

Top: House Finch, *Carpodacus mexicanus;* male. Bottom: House Finch; female.

House Sparrow

Passer domesticus   6 inches (15 cm)

The House Sparrow is a common bird near human habitations. It often can be seen around school yards, beach parks, picnic tables, and out-door restaurants—places where it can feed on the crumbs humans leave behind. It is occasionally found in agricultural areas, but rarely in the native forest. According to Andrew Berger the Hawaiian population is not as boldly colored and has paler legs than continental birds. The House Sparrow was imported from New Zealand about 1871.

Top: House Sparrow, *Passer domesticus;* male. Bottom: House Sparrow; female.

Hwamei

Garrulax canorus 9 inches (23 cm)

The *Hwamei* is also called the Chinese Thrush, Melodious Laughing-thrush, or Spectacled Thrush. It was a favorite cage bird of Chinese immigrants to Hawai'i and was reportedly shipped to Kaua'i from O'ahu in 1918. This introduced species is well established in both lowland and mountain habitats. Often heard but rarely seen, this species now thrives in areas in the native forest once occupied by Kaua'i's native thrushes, the *Kāma'o* and *Puaiohi*. The vocalizations of the *Hwamei* are loud repeated phrases of couplets and triplets.

Hwamei (Melodious Laughing-thrush), *Garrulax canorus.*

Japanese White-eye

Zosterops japonicus   4.5 inches (11 cm)

Brought to Hawai'i by the Hui Manu about 1927, the *Mejiro*, as it is called in Japan, is one of Kaua'i's most successful introductions. It is an extremely adaptable species that is equally common from the dry scrub of desertlike Mānā to the constantly wet forest near the Wai'ale'ale summit. The *Mejiro* travels in small, twittering flocks that become noisier when disturbed. In size and color, the Japanese White-eye can be mistaken for an endemic honeycreeper, but no Hawaiian honeycreeper has a white eye ring.

Japanese White-eye *(Mejiro), Zosterops japonicus.*

Java Sparrow

Padda oryzivora 6 inches (15 cm)

After the first unsuccessful introduction in 1865, the Java Sparrow was successfully reintroduced in the 1960s. A native of Indonesia, this is an attractive species that readily visits residential bird feeders. It can be seen from Princeville to Waimea and often feeds in small flocks mixed with Chestnut Mannikins. The grassy areas around the Kukui Grove Shopping Center and the Līhu'e airport are good places to find this colorful bird. Sexes do not differ in appearance, but juveniles are paler and lack the bold white patch on the cheek. The male Java Sparrow performs a curious courtship display to impress a potential mate. It alights next to the female, grips the branch tightly with its feet, and then rhythmically bounces up and down as if doing pushups.

Java Sparrow, *Padda oryzivora.*

Moa

Red Junglefowl

Gallus gallus   Male: 30 inches (76 cm); female: 17 inches (43 cm)

Brought by the Polynesians, the *Moa* or Red Junglefowl is Hawaiʻi's first introduced bird. The *Moa* is a bird mostly of the upland forest and is not to be confused with the many domestic chickens that roam the lowlands of Kauaʻi. As the name suggests, this species is red in contrast to the varied plumage of lowland birds. Interbreeding does occur, however. The best place to see the *Moa* is in the meadow fronting the Kōkeʻe Museum. They are present by the dozens and easily approached.

Moa (Red Junglefowl), *Gallus gallus;* male.

Moa; female.

Northern Cardinal

Cardinalis cardinalis 8–9 inches (20–23 cm)

Also known as the Kentucky Cardinal, Virginia Cardinal, or simply the Redbird, this attractive species can be found throughout the island. It consumes seeds avidly and is readily attracted to backyard feeders. Although the Northern Cardinal is present in the wet native forest, its numbers are much greater in the lowlands. In the morning and evening hours it can frequently be heard singing from high perches. Males are bright red and females are dull reddish brown. Juveniles resemble females, but have black bills. Large numbers of these birds are present in the *koa* and *kiawe* scrub in the Polihale area. This cardinal was imported from North America about 1929.

Northern Cardinal, *Cardinalis cardinalis.*

Northern Mockingbird

Mimus polyglottos   10 inches (25 cm)

Introduced by the Hui Manu in 1931 from North America, the mockingbird is common in dry habitats around the island. It easily can be seen in the area between Kalāheo and Mānā. This species often seeks a high perch like a telephone pole or power line from which it can sing its territorial repertoire. Mockingbirds on Kauaʻi are less vocal than those of the mainland, but do imitate the songs of other introduced birds. The mockingbird can even imitate the "meow" of a cat.

Northern Mockingbird, *Mimus polyglottos.*

Nutmeg Mannikin

Lonchura punctulata   4 inches (10 cm)

More frequently called the Ricebird or Spotted Munia, the Nutmeg
Mannikin was introduced from Southeast Asia in 1865. This bird was a
serious pest when rice farming was popular on Kaua'i. Flocks contain-
ing hundreds of birds would invade a field and destroy much of the
crop. Nutmeg Mannikins feed in the same manner as Chestnut
Mannikins but are more difficult to approach. Although it is seen mostly in lowland tall, green grass, this bird can also be found in the native
forest.

Nutmeg Mannikin, *Lonchura punctulata.*

Red-crested Cardinal

Paroaria coronata   7.5 inches (19 cm)

Often called the Brazilian Cardinal, this species is one of Kaua'i's most beautiful birds. It is a common sight around hotel grounds, golf courses, and parks, where it forages in pairs or small family groups. In areas high in human traffic it can be quite tame. Although this cardinal is found in greater numbers in the lowlands, it does exist at higher altitudes, occasionally visiting the meadow fronting the Kōke'e Museum. Fortunately, it rarely penetrates deeply into the native forest, where it might transport diseases or parasites. Males and females are similar. Juveniles differ in having a brown head and dark bill. Imported from South America about 1930, this species is only newly established. In 1960, it was not reported in a survey of the birds on Kaua'i.

Red-crested Cardinal, *Paroaria coronata.*

Ring-necked Pheasant

Phasianus colchicus Male: 33 inches (84 cm); female: 23 inches (58 cm)

This Asian species was one of the first game birds imported to Hawai'i.
It has been in the state since 1865 and is often referred to as the
Common Pheasant. It is a favorite of game bird hunters because of its
handsome plumage. It exists in two basic forms that frequently inter-
breed. On Kaua'i, the ring-necked form is, by far, the most common.
The green form is rare and probably exists in the wild only as released,
hand-raised birds. The plumage of the two forms is so different that tax-
onomists once considered them separate species. The Ring-necked
Pheasant inhabits open grassy areas or scrublike conditions and is rarely
seen in dense wet forest.

Top: Ring-necked Pheasant, *Phasianus colchicus* (ring-necked form). Bottom:
Ring-necked Pheasant (green form).

Rock Dove

Columba livia   12 inches (30 cm)

The Rock Dove, also known as the Common or Domesticated Pigeon, was one of the first birds introduced to Hawaiʻi after European "discovery." It is thought to have been imported from Europe about 1796. Nearly all populations of this species are closely associated with humans. In urban areas they roost on telephone and power lines or on the ledges of buildings. They often can be seen circling over populated areas in large flocks. Rock Doves express considerable variety in the color of their plumage. Pigeons, like parrots, are known to harbor psittacosis, a fungal disease that can be transmitted to humans.

Rock Dove, *Columba livia*.

Rose-ringed Parakeet

Psittacula krameri   16 inches (41 cm)

This escaped cage bird is the first parrot to become well established in Hawai'i. A yellow-green native of Central Africa and India, the Rose-ringed Parakeet can currently be observed on Kaua'i only in the area from Kōloa to Hanapēpē. A large flock of about a hundred birds is known to roost near the Kukuiolono Golf Course in Kalāheo. Other roosts have been reported in 'Ōma'o. An attractive species with a blue-green tail and striking red bill, the Rose-ringed Parakeet can do extensive damage to Kaua'i's seed, fruit, and vegetable crops. It will eat corn, lychee, papaya, and even bell peppers from the backyard garden. The species is considered a serious agriculture pest and an undesirable alien.

Rose-ringed Parakeet, *Psittacula krameri*.

Spotted Dove

Streptopelia chinensis   12 inches (30 cm)

Also called the Chinese Dove or Lace-necked Dove, this species was introduced from Asia in the nineteenth century. It is a common site on Kaua'i in both rural and populated areas. It is most easily approached in locations high in human traffic such as parking lots and school grounds. The Spotted Dove is the only dove that ventures into the Alaka'i Wilderness Preserve.

Spotted Dove, *Streptopelia chinensis.*

Warbling Silverbill

Lonchura malabarica   4.5 inches (11 cm)

The Warbling Silverbill is the least abundant of the three mannikins and often goes unnoticed. It is not known how the species was introduced to the Hawaiian Islands, but it was first recorded on the Big Island in the 1970s. On Kaua'i, it is most easily seen in dry areas, where it is attracted to water sources. The call of this small estrildid finch, which it utters in flight, has been described as a metallic "tic-tic-tic" sound like that of two coins being clicked together. It sometimes feeds in mixed flocks with Chestnut and Nutmeg Mannikins.

Warbling Silverbill, *Lonchura malabarica.*

Western Meadowlark

Sturnella neglecta   9.5 inches (24 cm)

As the name suggests, the meadowlark is a bird of open spaces and pasturelands. This gifted songster was introduced to Oʻahu and Kauaʻi in 1931, but only the Kauaʻi population has survived. To date, Kauaʻi remains the only place in the state where the Western Meadowlark can be seen. The species has a yellow breast accented with a prominent black **V**. In flight, white feathers are obvious on either side of the tail. The meadowlark likes to sing from high perches, so look for it atop fence posts in pastures from Hanapēpē to Hanalei. An excellent place to see this bird is along the north end of the Līhuʻe airport near the NOAA weather station.

Western Meadowlark, *Sturnella neglecta*.

White-rumped Shama

Copsychus malabaricus   10 inches (25 cm)

Released on Kaua'i in 1931 by the Hui Manu, the beautiful notes of this introduced thrush can now be heard island-wide. Although it sings at any time of day, the Shama Thrush, as it is sometimes called, is particularly vocal in the early morning and near sunset. Many consider the song of this species the most pleasing of all Kaua'i birdsongs. The shama is shy and not readily seen; however, it can occasionally be enticed into the open by imitating its call. Good places to see it are in the bamboo grove at Smith's Tropical Paradise and on the path that wanders along the Kaua'i Lagoons golf course. Like the *Hwamei,* it will frequently cross the road ahead of your vehicle on Kōke'e roads.

White-rumped Shama, *Copsychus malabaricus.*

Zebra Dove

Geopelia striata   8 inches (20 cm)

Data gathered in annual Christmas bird counts conducted by the
Hawai'i Audubon Society indicate that the Zebra or Barred Dove is one
of Kaua'i's most plentiful lowland birds. Although it does not frequent
the wet forest, it can be seen in the meadow fronting the Kōke'e
Museum. The Zebra Dove is a prolific breeder, nesting as many as five
times a year. Bowing courtship displays are a common sight. They can
become quite tame in populated areas, almost to the point of being
underfoot. The Zebra Dove was imported from Malaysia in 1922.

Zebra Dove, *Geopelia striata.*

Other Established Species Occasionally Seen or Heard

California Quail, *Callipepla californica*. The established population of this attractive species is frequently complemented by birds raised in captivity. Nevertheless, it remains difficult to find. This game bird prefers dry upland scrub.

Gray Francolin, *Francolinus pondicerianus*. Although common in dry *kiawe* habitats on the other main islands, this game bird is a rare find on Kaua'i.

Greater Necklaced Laughing-thrush, *Garrulax pectoralis*. At 13 inches (33 cm), this species is the largest of Kaua'i's thrushes. It forages in small groups that have been reported near 'Ōma'o and in Hulē'ia Stream valley.

Japanese Bush-Warbler, *Cettia diphone*. This secretive bird, known in Japan as the *Uguisu,* was first noticed on Kaua'i in the 1980s. It frequents dense underbrush, is gray, and measures from 5 to 6 inches (13–15 cm). Few people have been fortunate to get a glimpse of this bird. The Japanese Bush-Warbler is increasing on Kaua'i. It can often be heard at the scenic viewpoints in Hanalei, Waimea Canyon, and Kalalau; however, it is usually silent from September through December. The name *Uguisu* is an imitation of its song, and like the White-rumped Shama, the *Uguisu* responds well to attempts at mimicry.

Japanese Quail, *Coturnix japonica*. This small game bird is very difficult to flush and is therefore more often heard than seen. The species frequents the short-grass pastures in the Anahola area.

Red Avadavat, *Amandava amandava*. Also known as the Strawberry Finch, this rarely seen, escaped cage bird has recently become established on Kaua'i. Look for it mixed in with flocks of Chestnut and Nutmeg Mannikins. The species has been seen in the Kōloa area.

Red-billed Leothrix, *Leiothrix lutea*. Common in the forests of Maui and the Big Island, this colorful bird remains difficult to locate on Kaua'i. Birds have been reported along the trail to Hanakāpī'ai Valley.

Captive Birds

A survey of the birds of Kaua'i would not be complete without a final note on captive birds. There are several game bird breeders on the Garden Island who raise a variety of quails and pheasants. When mature, these birds are intentionally released into the wild in hopes of increasing the number of species available for hunting. For reasons not clearly understood, most do not thrive and fail to become established. The Wild Turkey and Northern Bobwhite that have been reported from a few localities around the island are both thought to fall under this nonestablished category.

A few hotels and tourist attaractions on Kaua'i also display captive species. A variety of introduced parrots, swans, domestic geese, and ducks can be seen.

Common Peafowl on the grounds of Smith's Tropical Paradise.

References

Anonymous. 1930a. Mrs. Isenberg Aids Movement to Get Birds. *Honolulu Advertiser,* 19 March, p. 9.

Anonymous. 1930b. Shipment of Rare Birds Coming Today. *Honolulu Advertiser,* 3 December, p. 4.

Anonymous. 1935. Board Permits Rare Birds to Enter at Last. *Honolulu Advertiser,* 6 December, p. 9.

Anonymous. 1936. Mrs. W. F. Dillingham Named President of Local Bird Club. Honolulu Advertiser, 22 December, p. 1.

Anonymous. 1960. Bird Thought Extinct Sighted in Kokee Area. *Honolulu Advertiser,* 28 July, p. A4.

Atkinson, Carter T., Robert J. Dusek, and William M. Iko. 1993. Avian Malaria Fatal to Juvenile 'I'iwi. *Hawaii's Forests and Wildlife* 8 (3): 1.

Berger, Andrew J. 1981. *Hawaiian Birdlife.* Honolulu: University of Hawai'i Press.

Carlquist, Sherwin. 1970. *Hawaii: A Natural History.* Garden City, New York: Natural History Press.

Culliney, John L. 1988. *Islands in a Far Sea: Nature and Man in Hawaii.* San Francisco: Sierra Club Books.

Denny, James H. 1996. A Craving for Nectar. *Spirit of Aloha* (April): 8–57.

Emerson, Nathaniel B. 1909. *Unwritten Literature of Hawaii, the Sacred Songs of the Hula, Collected and Translated, with Notes and an Account of the Hula.* Bureau of American Ethnology Bulletins, No. 38.

Foster, Nelson. 1993. *Bishop Museum and the Changing World of Hawai'i.* Honolulu: Bishop Museum Press.

Freed, Leonard A., Sheila Conant, and Robert C. Fleischer. 1994. Evolutionary Ecology and Radiation of Hawaiian Passerine Birds. Page 335 *in* E. Alison Kay, ed. *A Natural History of the Hawaiian Islands: Selected Readings,* 2nd ed. Honolulu: University of Hawai'i Press.

Grant, P. J. 1997. *Gulls: A Guide to Identification.* San Diego: Academic Press.

Harrison, Craig S. 1990. *Seabirds of Hawaii: Natural History and Conservation.* Ithaca, New York: Comstock Publishing Associates.

Hawai'i Audubon Society. 1993. *Hawaii's Birds.* Honolulu: Hawai'i Audubon Society.

Kimura, Bert Y., and Kenneth M. Nagata. 1980. *Hawaii's Vanishing Flora.* Honolulu: Oriental Publishing Company.

Munro, George C. 1944. *Birds of Hawaii.* Honolulu: Tongg Publishing Company.

National Geographic Society. 1992. *Field Guide to the Birds of North America,* 2nd ed. Washington: The National Geographic Society.

Olson, Storrs L. 1997. Address: Bird Fossil Findin.0gs and Evolution in Hawai'i. Hui O Laka volunteer appreciation luncheon. Kōke'e Museum, 10 May.

Olson, Storrs L., and Helen F. James. 1994. Descriptions of Thirty-two New Species of Birds from the Hawaiian Islands. Page 439 *in* E. Alison Kay, ed. *A Natural History of the Hawaiian Islands: Selected Readings,* 2nd ed. Honolulu: University of Hawai'i Press.

Perkins, R. C. L. 1903. *Fauna Hawaiiensis.* Cambridge, England: The University Press.

Pratt, H. Douglas. 1993. *Enjoying Birds in Hawaii: A Birdfinding Guide to the Fiftieth State.* Honolulu: Mutual Publishing Company.

Pratt, H. Douglas, Phillip L. Bruner, and Delwyn G. Berrett. 1987. *The Birds of Hawaii and the Tropical Pacific.* Princeton: Princeton University Press.

Pukui, Mary K., and Samuel H. Elbert. 1986. *Hawaiian Dictionary.* Honolulu: University of Hawai'i Press.

Pukui, Mary K., Samuel H. Elbert, and Esther T. Mookini. 1974. *Place Names of Hawaii.* Honolulu: University of Hawai'i Press.

Pyle, Robert L. 1997. Checklist of the Birds of Hawaii—1997. *'Elepaio* 57:129–138.

Richardson, F., and J. Bowles. 1964. *A Survey of the Birds of Kauai, Hawaii.* Honolulu: Bishop Museum Press.

Scott, Michael J., S. Mountainspring, F. L. Ramsey, and C. B. Kepler. 1986. *Forest Bird Communities of the Hawaiian Islands: Their Dynamics, Ecology, and Conservation.* Studies in Avian Biology No. 9. Lawrence, Kansas: Allen Press, Inc.

Tenbruggencate, Jan. 1997. Birds Thrive in Forest Free of Rats. *Honolulu Advertiser,* 7 July.

Wagner, Warren L., Derral R. Herbst, and S. H. Sohmer. 1990. *Manual of the Flowering Plants of Hawai'i.* Honolulu: University of Hawai'i Press and Bishop Museum Press.

Wilson, Scott B., and Arthur H. Evans. 1890–1899. *Aves Hawaiiensis: The Birds of the Sandwich Islands.* London: R. H. Porter.

A Checklist of the Birds of Kaua'i

Species	Date	Place
Native Forest Birds		
'Akeke'e (Kaua'i *'Ākepa*), *Loxops caeruleirostris*		
'Akikiki (Kaua'i Creeper), *Oreomystis bairdi*		
'Anianiau, Hemignathus parvus		
'Apapane, Himatione sanquinea sanquinea		
I'iwi, Vestiaria coccinea		
Kāma'o (Large Kaua'i Thrush), *Myadestes myadestinus* *		
Kaua'i *'Akialoa, Hemignathus ellisianus procerus* *		
Kaua'i *'Amakihi, Hemignathus kauaiensis*		
Kaua'i *'Elepaio, Chasiempis sandwichensis sclateri*		
Kaua'i *Nukupu'u, Hemignathus lucidus hanapepe* *		
'Ō'ō 'ā'ā, Moho braccatus *		
'Ō'ū, Psittirostra psittacea *		
Puaiohi (Small Kaua'i Thrush), *Myadestes palmeri* *		
Pueo (Hawaiian Owl), *Asio flammeus sandwichensis*		

* denotes a species officially listed as "endangered" by the U.S. Fish and Wildlife Service.

Wetland Birds and Shore Birds

Species	Date	Place
Aeʻo (Hawaiian Stilt), *Himantopus mexicanus knudseni* *		
ʻAkekeke (Ruddy Turnstone), *Arenaria interpres*		
ʻAlae keʻokeʻo (Hawaiian Coot), *Fulica alai* *		
ʻAlae ʻula (Hawaiian Gallinule), *Gallinula chloropus sandvicensis* *		
American Wigeon, *Anas americana*		
ʻAukuʻu (Black-crowned Night-Heron), *Nycticorax nycticorax hoactli*		
Green-winged Teal, *Anas crecca*		
Hunakai (Sanderling), *Calidris alba*		
Kioea (Bristle-thighed Curlew), *Numenius tahitiensis*		
Kōlea (Pacific Golden-Plover), *Pluvialis fulva*		
Koloa maoli (Hawaiian Duck), *Anas wyvilliana* *		
Koloa māpu (Northern Pintail), *Anas acuta*		
Koloa mohā (Northern Shoveler), *Anas clypeata*		
Lesser Scaup, *Aythya affinis*		
Long-billed Dowitcher, *Limnodromus scolopaceus*		
Mallard, *Anas platyrhynchos*		
Nēnē (Hawaiian Goose), *Branta sandvicensis* *		
Pectoral Sandpiper, *Calidris melanotos*		
Ring-necked Duck, *Aythya collaris*		
ʻŪlili (Wandering Tattler), *Heteroscelus incanus*		

* denotes a species officially listed as "endangered" by the U.S. Fish and Wildlife Service.

SPECIES	DATE	PLACE

Seabirds

SPECIES	DATE	PLACE
'Ā (Brown Booby), *Sula leucogaster plotus*		
'Ā (Red-footed Booby), *Sula sula rubripes*		
'Akē'akē (Band-rumped Storm-Petrel), *Oceanodroma castro*		
'A'o (Newell Shearwater), *Puffinus newelli* *		
Black-footed Albatross, *Phoebastria nigripes*		
Bonaparte Gull, *Larus philadelphia*		
Bulwer Petrel, *Bulweria bulwerii*		
Christmas Shearwater, *Puffinus nativitatis*		
'Ewa'ewa (Sooty Tern), *Sterna fuscata oahuensis*		
Glaucous-winged Gull, *Larus glaucescens*		
'Iwa (Great Frigatebird), *Fregata minor palmerstoni*		
Koa'e kea (White-tailed Tropicbird), *Phaethon lepturus dorotheae*		
Koa'e 'ula (Red-tailed Tropicbird), *Phaethon rubricauda rothschildi*		
Laughing Gull, *Larus atricilla*		
Mōlī (Laysan Albatross), *Phoebastria immutabilis*		
Noio (Black Noddy), *Anous minutus melanogenys*		
Noio kōhā (Brown Noddy), *Anous stolidus pileatus*		
Pākalakala (Gray-backed Tern), *Sterna lunata*		
Ring-billed Gull, *Larus delawarensis*		
'Ua'u (Dark-rumped Petrel), *Pterodroma phaeopygia sandwichensis* *		
'Ua'u kani (Wedge-tailed Shearwater), *Puffinus pacificus chlororhynchus*		

* denotes a species officially listed as "endangered" by the U.S. Fish and Wildlife Service.

SPECIES	DATE	PLACE

Introduced Birds

SPECIES	DATE	PLACE
Barn Owl, *Tyto alba*		
Black Francolin, *Francolinus francolinus*		
California Quail, *Callipepla californica*		
Cattle Egret, *Bubulcus ibis*		
Chestnut Mannikin, *Lonchura malacca*		
Chukar, *Alectoris chukar*		
Common Myna, *Acridotheres tristis*		
Domestic Chicken, *Gallus domesticus*		
Erckel Francolin, *Francolinus erckelii*		
Gray Francolin, *Francolinus pondicerianus*		
Greater Necklaced Laughing-thrush, *Garrulax pectoralis*		
House Finch, *Carpodacus mexicanus*		
House Sparrow, *Passer domesticus*		
Hwamei (Melodious Laughing-thrush), *Garrulax canorus*		
Japanese Bush-Warbler, *Cettia diphone*		
Japanese Quail, *Coturnix japonica*		
Japanese White-eye *(Mejiro), Zosterops japonicus*		
Java Sparrow, *Padda oryzivora*		
Moa (Red Junglefowl), *Gallus gallus*		
Northern Cardinal, *Cardinalis cardinalis*		
Northern Mockingbird, *Mimus polyglottos*		
Nutmeg Mannikin, *Lonchura punctulata*		
Red Avadavat, *Amandava amandava*		
Red-billed Leothrix, *Leiothrix lutea*		

(continued)

(continued)

SPECIES	DATE	PLACE
Red-crested Cardinal, *Paroaria coronata*		
Ring-necked Pheasant, *Phasianus colchicus*		
Rock Dove, *Columba livia*		
Rose-ringed Parakeet, *Psittacula krameri*		
Spotted Dove, *Streptopelia chinensis*		
Warbling Silverbill, *Lonchura malabarica*		
Western Meadowlark, *Sturnella neglecta*		
White-rumped Shama, *Copsychus malabaricus*		
Zebra Dove, *Geopelia striata*		

A complete listing of all of Hawai'i's endangered species can be found on the internet at: http://www.fws.gov/~r9endspp/statl-hi.html#LnkHI

Space has been provided for the reader to record any additional species he or she might encounter on Kaua'i. I would appreciate a notification of the sighting of any species not mentioned in this book. Please contact Jim Denny, P.O. Box 232, Kekaha, HI 96752 (telephone: 808–337–1081; e-mail: jhdenny@aloha.net).

Index

The index includes taxonomic and vernacular names. Hawaiian and scientific names are listed in *italics*. Page numbers in **boldface** designate the location where a species is illustrated.

About the Author

JIM DENNY received his B.A. degree at Louisiana Tech and his B.S. at the University of Hawai'i at Mānoa. He is employed at Kaua'i Veterans Memorial Hospital as a medical technologist. In 1968, when Jim moved from the bayous of Louisiana to Hawai'i, he brought with him his love for the outdoors. In the forests of Kaua'i, he developed a keen interest in the endemic flora and fauna of the Islands. Through his writing and photography Jim hopes to convey to others the beauty and plight of Kaua'i's native species. His articles and photos have been published locally in *Island Scene Magazine,* in the award-winning *Kaua'i Magazine,* in Aloha Airline's in-flight magazine, *Spirit of Aloha,* and on the front pages of the *Garden Island* and *Kaua'i Times* newspapers. Nationally, his photos have appeared in the literature of the U.S. Fish and Wildlife Service, in publications of private conservation groups, in biology textbooks, in *Audubon, Smithsonian,* and *National Geographic* magazines, as well as on the *Discovery Channel Online.* Internationally, his work has appeared on Japanese television and in the acclaimed nature magazine SINRA. In 1991, Jim became one of only a few to ever photograph the *Puaiohi,* a critically endangered species residing in the Alaka'i Swamp. His work has been exhibited in the Hall of Discovery at the Bishop Museum and at the Kōke'e Museum. In 1994, Jim was honored as the "volunteer of the year" by the environmental organization Hui O Laka for his efforts to educate the public about Hawaiian birds. Jim frequently lectures at local schools and to visiting eco-tour groups about Kaua'i's precious birds.